THE DEVIL DOG DICTIONARY

THE DEVIL DOG DICTIONARY

JOHN BURKE

STRIKEPOINT
PUBLISHING

Copyright © 2025 by John Burke
All rights reserved.

Published by Strikepoint Publishing
www.strikepoint-publishing.com

No part of this book may be reproduced in any manner whatsoever without written permission except in the case of brief quotations embodied in critical articles and reviews.

Neither the United States Marine Corps nor any other component of the Department of Defense has approved, endorsed, or authorized this book.

ISBN: 979-8-9998133-0-5

First Printing, 2025

"Retreat, hell! We just got here."

— Captain Lloyd W. Williams, USMC
Battle of Belleau Wood, June 1918

CONTENTS

Dedication   v
Preface   ix
Preserving the Language of the U.S. Marine Corps   xi

- A -   1
- B -   3
- C -   10
- D -   15
- E -   19
- F -   21
- G -   25
- H -   29
- I -   32
- J -   34
- K -   36
- L -   38
- M -   41
- N -   45
- O -   47
- P -   49

- Q -	52
- R -	53
- S -	55
- T -	62
- U -	65
- V -	67
- W -	68
- X -	70
- Y -	71
- Z -	72
Numeric Terms	73
Appendix 1: Pronunciation Guide	75
Appendix 2: Marine-to-Civilian Decoder Chart	77
Appendix 3: Naval Terms Every Marine Should Know	79
Appendix 4: NATO Phonetic Alphabet	82
Appendix 5: Radio Brevity Codes	84
Appendix 6: Marine Corps Ranks & Insignia	87
Appendix 7: Key Marine Corps Dates & Battles	94
Appendix 8: Historic Quotes From and About Marines	97
Appendix 9: Marine Corps Values	102
Appendix 10: The Creed of the U.S. Marine	103
About the Author	105

Preface

You step off the bus at zero-dark-thirty and step onto the yellow footprints at Parris Island (or San Diego if you're a Hollywood Marine). You're disoriented and a bit anxious — but that's when the transformation begins for enlisted Marines.

Over the next 13 weeks, the Corps tears down the civilian in you and builds something tougher. You learn the drill, the discipline and — maybe without even realizing it at first — you start to pick up the language. It's in the cadence calls, the way your drill instructors bark out orders, the periods of instruction where you learn the history, customs and traditions of the Corps, how to use your gear, wear your uniform and master your rifle.

Pretty soon you're tossing around words like "scuttlebutt," "squared away" and "unsat" without a second thought. By the time you hit the fleet, that language is part of you — a badge of belonging that marks you as a Marine just as surely as your high-and-tight or your Eagle, Globe and Anchor.

It happened to me and countless other Marines over the past 250 years. From Tun Tavern to Belleau Wood, Iwo Jima and the Frozen Chosin to the jungles of Vietnam, the streets of Fallujah and the poppy fields of Helmand — Marines have faced determined adversaries and met them with ferocity. Out of that hardship, pride and camaraderie, a vast vocabulary has grown. It's more than slang; it's how Marines make sense of the world, cope with hardship and keep each other laughing no matter how deep in the suck they are.

This little book is my way of capturing that living language. It's a dictionary, sure, but it's also a snapshot of who we are — tough, loyal, dark-humored and always ready to improvise, adapt and overcome. These words stitch together Marines from every generation. Whether you wore the uniform decades ago, more recently or today, we all still speak the same way.

"Once a Marine, Always a Marine."

That's not just a slogan on a bumper sticker. It's a lifelong ticket into a tightly knit culture that stretches from 10 Nov. 1775 to today. And the way we talk — well, that's part of what keeps that bond so strong.

I put this together for Marines, families, history buffs and anyone curious about what makes us tick. May it preserve a little piece of our Corps, so future generations know not just how Marines fight, but how we communicate, how we think, how we bust each other's chops and how we keep going beyond when others would quit.

If the walls of the National Museum of the Marine Corps could talk, you can bet they'd echo a lot of the words you'll find in these pages.

Semper Fi,

John Burke
Sergeant of Marines (1983–1989)

Preserving the Language of the U.S. Marine Corps

Acknowledgments and a Call to Action

The language of the United States Marine Corps is not fixed in time — it is a living tradition, forged in boot camp, Officer Candidate School (OCS) and The Basic School (TBS) and passed from generation to generation for two-and-a-half centuries. With every campaign, every deployment and every fresh boot stepping off the yellow footprints, our lexicon grows. It is shaped by experience, steeped in history and sharpened by the unique humor and grit that define the Corps.

This volume does not attempt to catalog every piece of military jargon or every expression used across the branches. Nor does it claim to include every term spoken across every Marine Corps duty station. Instead, this dictionary aims to focus on the words, phrases and expressions that are distinctly Marine — terms we originated or adopted and made unmistakably our own.

Many of the entries contained here come from firsthand experience during my time in the Corps. But no work of this scope is built on memory alone. I also relied on a range of sources — from printed references to online glossaries, from official publications to informal forums — to ensure accuracy and depth.

In particular, I am indebted to the **Marine Corps University** and the **official publications of the United States Marine Corps**, including doctrinal materials and public resources such as the Marine Corps Core Values, the Rifleman's Creed and historic collections of Marine quotes. Because these works are produced by the U.S. Government, they are in the public domain — but the duty of credit still stands.

I am also grateful to the Marines past and present, and to the historians, writers, and fellow Devil Dogs whose work, words and wit helped shape this collection.

But this dictionary is not meant to be final. Like the Corps itself, it is always evolving — and that's where you come in.

If you know of a term we've missed, a phrase from today's fleet or some salty slang that deserves to be remembered, **visit devildogdictionary.com** and share it. Submissions and feedback are welcome and will be considered for inclusion in future editions.

Because as long as there are Marines, there will be new ways to say what only Marines can truly understand.

- A -

Aboard (adverb): On base, on ship or inside unit area.

ACE (acronym): Aviation Combat Element.

Adapt and overcome (phrase / motto): Marine mantra to handle anything.

Admin (noun): Clerical work or the office personnel who handle it.

Admin sep (noun): A process where a Marine can be involuntarily discharged form service, typically due to misconduct or unsatisfactory performance.

Aft (adverb): Toward the stern of a ship.

Air Krulaks (noun): Black, Gore-Tex-lined combat boots issued from 1997 to 2001, named after the 31st Commandant of the Marine Corps.

A.J. Squared Away (adjective): A Marine who is perfectly squared away — everything in order. The opposite of a shitbird.

All hands (noun): Everyone in the unit; entire group.

All hands on deck (phrase / idiom): Everyone in a unit or command should assist in a particular task or situation.

Alphas (noun): Marine Corps Service Alpha uniform (green coat & trousers, khaki shirt & tie).

Amidships (adverb): In or toward the middle part of a ship or aircraft; midway between the ends.

Article 15 (noun): Non-judicial punishment under the UCMJ; called NJP in the Corps.

Ashore (adverb): On the shore; on land rather than at sea or on the water.

Ass pack (noun): Extra gear or secondary pack lashed onto the main pack.

Assault pack (noun): Small day pack for combat essentials.

Asshole to Belly Button (expression / idiom): Standing close together in a line.

As you were (interjection): Resume former activity.

Attention on Deck (phrase): Verbal command used to alert personnel that a senior officer or dignitary is entering the area, signaling them to come to attention.

Aye aye (interjection): Acknowledgement of a verbal order, signifying the order will be carried out.

- B -

Baby on board (expression): Sarcastic remark for a brand-new Marine arriving at the unit.

Back brief (verb): Junior Marines repeat instructions to ensure understanding.

Back in the rear with the gear (expression): Refers to support troops who are not on the front line.

Back on the block (expression): Acting like a civilian.

Bag nasty (noun): Bagged meal issued to Marines in boot camp or the field.

Bail out (verb): Exit quickly.

BAMCIS (acronym): Six troop leading steps: **B**egin planning, **A**rrange reconnaissance, **M**ake reconnaissance, **C**omplete the planning, **I**ssue the order, **S**upervise.

Banana Wars (noun): Early 20th-century interventions in Central America and the Caribbean.

Bang bang (noun): Rifle or small arms fire.

Bang box (noun): Ammunition box.

Bang out (verb): Typically refers to performing physical exercises, particularly pull-ups.

Barney-style (adverbial phrase / expression): To perform strictly according to regulation; idiot-proof; simplified for the benefit of mental underachievers.

Barracks Bunny (noun): Female Marines that tend or are rumored to sleep around in the barracks

Barracks Cover (noun): Garrison (frame) cap.

Barracks Lawyer (noun): Marine who gives unsolicited or incorrect legal advice.

Barracks Rat (noun): Marine who never leaves base, often due to lack of money.

Base plate (noun): Bottom piece of a mortar system.

Battery Operated Grunt (noun): Combat radio operator.

Battle buddy (noun): Assigned partner for accountability.

Battle rattle (noun): Full combat gear.

Battle Sight Zero or BZO (verb): To adjust weapon sights to establish a standard zero before making adjustments for windage or elevation.

Battle space (noun): The operational environment, encompassing all aspects of the physical, informational and cognitive domains that influence military operations.

Bayonet course (noun): Obstacle course for bayonet training.

BBD (noun): Bigger, Better Deal. Someone who left for a better opportunity.

BCGs (noun): Boot camp-issued glasses, jokingly called Birth Control Glasses.

BDA (noun): Battle Damage Assessment.

Beans and bullets (noun): Supplies of food and ammunition.

Beans, bullets and bandages (expression): Food, ammo and medical supplies — essentials of logistics.

Beat your face (command / imperative): Drop and do push-ups.

Beer light (noun): Symbolically turns green when operation ends, time for liberty.

Belay (verb): To make fast or secure, as in "belay the line."

Belay that (interjection): Cancel or disregard the previous statement or order.

Belleau Wood (noun): WWI battle where Marines earned the "Devil Dog" nickname.

Belt fed (noun/adjective): Machine gun, or jokingly someone very intense.

Belt fed grunt (noun): Machine gunner.

Benotz (noun): Generic term for an unnamed junior Marine, often one who gets into trouble.

BEQ (noun): Bachelor Enlisted Quarters. Living spaces that often have shared or semi-private bathrooms and may include common areas like laundry facilities, recreation rooms and sometimes even dining or vending areas

BFT (noun): Blue Force Tracker. GPS tracking system for friendly forces.

Billet (noun): Assignment or job; place of residence.

Big Marine Corps (noun): The institution as a whole, often used when discussing top-down orders.

Bird (noun): Any aircraft.

Bird colonel (noun): Colloquial term for a Colonel (O-6). This term differentiates them from Lieutenant Colonels, who are also sometimes referred to as "Colonel."

Bird farm (noun): Aircraft carrier.

Birthday ball (noun): Formal celebration of the Corps '10 November birthday.

Birthday message (noun): Annual address from the Commandant of the Marine Corps on 10 November.

Black Cadillacs (noun): Combat boots.

Black gear (noun): Can refer to unencrypted radios, distinct from the "green gear" of encrypted radios.

Blood pinning (verb): Hazing ritual in which senior Marines pound a newly earned rank insignia into the recipient's chest, causing bleeding.

Blood stripe (noun): Scarlet red stripe worn on dress blue trousers, symbolizing heavy Marine losses at the Battle of Chapultepec in 1847.

Blouse (noun): Cammie shirt.

Blow chow (verb): To vomit, especially during intense PT.

Blowout kit (noun): Individual trauma kit.

BLT (acronym): Battalion Landing Team.

Blue Falcon (noun): Someone who betrays or screws over fellow Marines; short for "Buddy Fucker."

Blue on blue (noun): Friendly fire incident.

Blues (noun): Marine Corps Dress Blue uniform.

Blues and twos (noun): Dress blues worn with medals instead of ribbons.

Body bag PT (noun): Extremely grueling physical training.

Bone yard (noun): Area for old or broken vehicles.

Booger hook (noun): Trigger finger, keep it off the trigger.

Boom boom room (noun): Barracks room used for personal "rendezvous."

Boomerang (noun): A Marine who gets out and then later re-enlists.

Boonie cover (noun): Wide-brim field cover.

Boot (noun): A new Marine fresh from recruit training.

Boot bands (noun): Elastic bands used to "blouse" trousers over boots.

Boot camp (noun): Marine Corps Recruit Training.

Boot drop (noun): New batch of Marines arriving at unit.

Boot Lewy (noun): Nickname for a newly commissioned 2nd Lt. See "Butter Bar."

Boot sock (noun): Thick green or black sock issued with boots.

Boots and Utes (expression): Physical training conducted in boots and camouflage utilities.

BOQ (noun): Bachelor Officer Quarters.

Box of Grid Squares (noun): Hazing joke to send boots to "find."

Brain housing group (noun): Skull or head.

Brass (noun): Officers, or spent shell casings.

Bravo Zulu or BZ (expression): Navy signal for "well done."

Bravos (noun): Same as the service "A" uniform except that the service coat is not worn. The long-sleeve khaki shirt is worn with the appropriate necktie and necktie clasp/necktab.

Break it down Barney style (expression): Explain it very simply, step by step.

Brig (noun): Military jail.

Brig rat (noun): Someone who spends time in the brig.

Broke-dick (adjective): Description of a Marine who underperforms or unreliable gear (also slang for gear that fails).

Brown Bagger (noun): Married Marine.

Brown side out (expression): Refers to old reversible field cammies, specifying which side (green or brown) is worn outward.

Buddy fucker (noun): Disloyal Marine; also see "Blue Falcon."

Buddy rush (verb): Marine fire team movement technique.

Bulkhead (noun): Wall.

Bum scoop (noun): Bad or incorrect info.

Busted (adjective): Reduced in rank by NJP.

Butter bar (noun): Nickname for a 2nd Lt. See "Boot Lewy."

- C -

Cake cutting ceremony (noun): Marine Corps birthday tradition.

Cammie house (noun): Barracks.

Cammie stick (noun): Camouflage face paint.

Cammies (noun): The Marine Corps Combat Utility Uniform (MCCUU).

Cammie up (verb): Apply camo.

Camp Wilson (noun): Live-fire training area at Marine Corps Air Ground Combat Center, 29 Palms.

Cank'd (verb): Cancelled.

Canon cocker (noun): Artillery Marine.

Cans (noun): Ear protection.

Cap Marine (noun): Retired Marine.

Captain's Mast (noun): Term used for Non-Judicial Punishment (NJP), also known as "Office Hours."

Cardio hill (noun): Base running landmark.

Carry on (interjection): Order to continue work or duties.

Casualty feeder report (noun): Casualty report format used by Marines.

Cat hole (noun): Hole dug for burying human waste.

Cattle Car (noun): Vehicle to move recruits or Marines en masse.

CAX (noun): Combined Arms Exercise.

CE (acronym): Combat Element.

CFT (noun): Combat Fitness Test.

Charger / Hard Charger (noun): Aggressive Marine.

Charlie Mike (expression): "Continue Mission," used in Marine comms.

Charlies (noun): The short-sleeve khaki shirt with appropriate service trousers or skirt/slacks is designated as the service "C" uniform. During the winter season, commanders may, at their discretion when the weather requires, authorize the service "C" uniform.

Check gear (verb): Inspect equipment.

Check in (verb): Report on arrival.

Cherry blast (noun): New Marines who overly decorate their pack.

Chesty (noun): Nickname for LtGen Lewis "Chesty" Puller.

Chesty Puller (noun): Most decorated Marine in history, known for toughness and combat leadership.

Chesty would be proud (expression): The ultimate Marine endorsement.

China Marine (noun): Pre-WWII Marines in China.

Chinstrap (noun): New Marine, implying they're so fresh they still have their chinstrap hanging down.

Chit (noun): Written authorization or receipt.

Chop (verb): To formally report to a new duty station or unit.

Chosin Few (noun): Veterans of Chosin Reservoir.

Chosin Reservoir (noun): Site of the legendary breakout by surrounded Marines in Korea, winter 1950.

Chow (noun): Food.

Chow Hall (noun): A mess hall or dining facility.

Chow line (noun): Food line.

Christmas tree (noun): Overloaded Marine.

Clacker (noun): Detonator for Claymores.

Clear hot (expression): Cleared to fire.

Click (noun): Kilometer. Also spelled "klick."

Close air support (noun): A USMC innovation where aircraft provide direct, close-in fire and maneuver support to ground forces.

Cluster fuck / Charlie Foxtrot (noun): Complete mess, while used across services, it's deeply embedded in Marine humor.

CO (noun): Commanding Officer.

Cobra (noun): AH-1, specifically Marine helo gunship.

Colors (noun/verb): The flag. As a verb, raising or lowering the flag.

Colors ceremony (noun): Raising/lowering flag.

Combat camera (noun): Photographers or videographers embedded with units.

Combat cargo (noun): Marines on ship duties.

Comms (noun): Communications, comms equipment or the discipline itself.

Comms check (noun): Radio verification

Com-rats: (noun): Commuted rations; pay for meals off base.

Contact left/right/front/rear (expression): Marine contact calls.

CONUS (noun): Continental United States.

Corpsman (noun): Navy medic assigned to Marines.

Cover (noun): Hat.

Cover and concealment (noun): Military tactic — "cover" protects from fire, "concealment" hides from view.

Cover down (verb): Align in formation.

Cover your six (expression): Watch your back.

Coyote brown (noun): Standard Marine gear color.

Crayon eater (adjective): Modern self-deprecating joke among Marines implying low intelligence.

Crucible (noun): Final training event in Marine Corps boot camp, symbolizing the transformation from recruit to Marine.

CS gas (noun): Tear gas.

CSSE (acronym): Combat Service Support Element.

- D -

Dark Green / Light Green (adjective): Informal reference to a Marine's skin tone; reflects the saying that Marines are not black or white, just different shades of green.

Dead horse (noun): Unpaid obligation.

Deadlined (adjective): Gear unserviceable.

Death before dishonor (expression): Core warrior ethos — losing honor is worse than death.

Death blossom (noun): Panicked firing.

Deck (noun): Floor.

Deltas (noun): The blue dress uniform with short-sleeve khaki shirt (without coat).

Deployment (noun): Temporarily moved from permanent duty station, generally to an overseas location or on a ship.

Det (abbreviation): Detachment.

Deuce gear (noun): USMC individual field gear.

Deuce-and-a-half (noun): 2.5 ton truck.

Devil Doc (noun): Respectful term for Navy corpsman with Marines.

Devil Dog (noun): Nickname for Marines, said to originate from German "Teufel Hunden" at WWI Battle of Belleau Wood.

Devil Pup (noun): Marine child or junior Marine.

DI (noun): Drill Instructor.

DI hut (noun): Drill Instructors' office in squad bay.

DI shuffle (noun): Distinct DI march.

Diddy bag (noun): Personal bag for essentials.

Diddy bop (adjective): Marching in a sloppy or non-military manner.

Direct reflection of leadership (adjective): Used sarcastically to rib an NCO when one of his or her Marines gets in trouble.

Dirt det (noun): Can consist of any number of Marines and equipment that are forward-deployed from their primary base of operations.

Dirty name tape (noun): Faded uniform tag.

Doggie (noun): Nickname for U.S. Army soldier.

Dog tags (noun): ID tags.

Don't pop smoke yet (expression): Don't leave yet.

Donkey dick (noun): Slang for siphon tube or nozzle.

Door kicker (noun): Infantry Marine / breacher.

Double feed (noun): Weapon jam.

Double tap (verb): Two rapid shots.

Double time (verb): Jog.

Down range (noun/adverb): Toward the target area or deployed in a combat zone.

Dragon (noun): USMC shoulder-fired anti-armor guided missile system.

Dress Blues (noun): Marine's iconic uniform.

Dress right, dress (command): Align in formation.

Drill card (noun): Card for DIs.

Drill Instructor / DI (noun): A specially selected and trained non-commissioned or staff non-commissioned officer who is responsible for transforming civilians into Marines.

Drop (verb): Fail out.

Drop gear (verb): Take off pack.

Drop on request (verb): Voluntarily quit a tough course (DOR).

Drop your cocks and grab your socks (expression): Vulgar for hurry-up.

Drop your pack (verb): Relax.

Dry fire (verb): Practice weapon firing without live ammo.

Dry hole (noun): Searched area found empty.

Duck walk (verb): Squat walk used in PT or recruit training.

Ductus Exemplo (expression / motto): Latin term that means "lead by example." Official motto of Officer Candidates School (OCS).

Dummy cord (noun): Cord used to tie gear to body so it doesn't get lost.

- E -

E-tool (noun): Entrenching tool.

Eagle, Globe and Anchor / EGA (noun): USMC emblem.

EAS (noun): End of Active Service.

Eat dirt (verb): Get down under fire.

Eat the apple, fuck the Corps (expression): Marriage means risking your career.

Echelon (noun): Levels of maintenance and support within a unit, not a specific unit or organization itself.

Echo Tango Sierra (expression): Eat This Shit. NATO phonetic spelling slang.

E-Club (noun): Enlisted club.

EGA (noun): Eagle, Globe and Anchor.

Eight-point cover (noun): Marine utility cap.

Embark (verb): Load on ship/aircraft.

Embassy duty (noun): Marine Security Guard (MSG) assignments.

Embrace the suck (expression): To accept and persevere through challenging, uncomfortable or unpleasant situations.

Engage (verb): Open fire.

Ensign (noun): National flag or naval officer rank.

Esprit de Corps (noun): A sense of pride, fellowship and common loyalty.

Evening Colors (noun): Flag lowering ceremony.

Every Marine a rifleman (expression): Core doctrine that every Marine, regardless of MOS, is trained to fight as infantry.

Eye pro (noun): Protective eyewear.

Eyes right! (command): Command given to signal a formal display of respect, often during parades or ceremonies, by having troops turn their heads and eyes to the right while maintaining forward march.

- F -

Fallujah (noun): USMC's iconic 2004 urban fight in Iraq.

Fangs (noun): Eating utensils, usually in the field.

Fangs out (expression): Aggressive.

Fartsack (noun): Sleeping bag.

Field Day (noun): Thorough cleaning of barracks or offices, traditionally on Thursday.

Field expedient (adjective): Improvised from available materials.

Field jacket (noun): Cold weather coat.

Field op (noun): Field training.

Field scarf (noun): Neck tie.

Field shower (noun): Water bottle rinse.

Field strip (verb): Disassemble weapon.

Fifty-cal (noun): M2 .50 machine gun.

Fighting hole (noun): Two-man dug in position.

Fire for effect (command): Full artillery barrage.

Fire team (noun): Four-man unit.

Fire Watch (noun): Guard duty.

Fire Watch ribbon (noun): National Defense Medal, referred to jokingly because recruits earn the medal during wartime.

Firing line (noun): Rifle line.

First to fight (expression): Historic slogan highlighting the Marine Corps' rapid response.

FITREP (Fitness Report) (noun): Evaluation report written on Marines (Sergeant and above) detailing proficiency and conduct and fitness for command, reviewed for promotion.

Fitty (noun): .50 cal slang.

Five jump chump (noun): Teasing term for Airborne-qualified Marines with minimum jumps.

Five paragraph order (noun): Marine format for operation orders — SMEAC (**S**ituation, **M**ission, **E**xecution, **A**dmin & Logistics, **C**ommand & Signal). See SMEAC entry.

Flat top (noun): Standard "high & tight" haircut variation.

Fleet Marine Force (FMF): (noun): A balanced force of combined arms comprising land, air and service elements of the Marine Corp

Flex (verb): Adjust plans.

Float (noun): Deployment by ship.

Flex cuff (noun): Zip-tie handcuffs.

Flush (verb): Drive enemy out.

FNG (noun): Fucking New Guy.

FOB (noun): Forward Operating Base.

Fobbit (noun): Stays at FOB.

Force multiplier (noun): Anything that enhances the effectiveness of Marine Corps operations, allowing them to achieve more with the same or fewer resources. This can include technology, tactics, partnerships and even intangible factors such as esprit de corps.

Force recon (noun): Elite recon unit.

Fore / Forward (noun): The front or forward part of a ship.

Formation (noun): Unit lineup.

Forward observer (FO): (noun): Calls indirect fire.

Foul deck (noun): Unsafe for aircraft landing.

FPF (noun): Final Protective Fire

Frag (noun): Grenade.

FRAGO (noun): Fragmentary order.

French Fourragère (noun): Green & red braided cord worn by Marines of 5th & 6th Marine Regiments, honoring WWI valor at Belleau Wood, Soissons & Champagne.

Frog sticker (noun): Bayonet.

Frozen Chosen (noun): Playful jab at Parris Island grads (mirrors "Frozen Chosin").

Frozen Chosin (nickname): Chosin Reservoir.

FUBIJAR (expression): Fucked Up, But I'm Just A Reservist.

Full battle rattle (noun): Complete USMC gear.

- G -

Gaff off (verb): To disregard or ignore a command or person, often in a disrespectful or insubordinate way.

Gaggle / Gagglefuck (noun): Disorganized group, boot camp favorite.

Galley (noun): Kitchen or area with kitchen on a ship.

Gangway (noun): A passageway, opening in the railing or bulwark of a ship.

Gang way (interjection): Stand back. Move aside.

Garrison (noun): Military base or installation.

Gas chamber (noun): Training facility for CS tear gas.

GCE (acronym): Ground Combat Element.

Gear adrift (expression): Gear left unsecured, inviting trouble.

Gear check (noun): Equipment inspection.

Gear locker (noun): Storage room.

Geedunk (noun): Junk food, snacks. See "Pogey bait."

Get some! (interjection): Iconic Marine battle cry.

Get your shit wired (expression): Get organized; get yourself squared away.

Ghost Turds (noun): Dust balls on the deck, often joked about in field day inspections.

Gig (noun): Minor inspection fail.

Gig line (noun): Alignment of shirt, belt buckle, fly.

Giz (noun): General issue item (GI).

Glamor shots (noun): Deployment pics.

Glow belt (noun): PT reflective belt.

Go fasters (noun): Running shoes.

Goat rope (noun): A chaotic, disorganized and confusing situation.

Good initiative, bad judgment (expression): When a Marine does something for a good reason, but it ends in disaster.

Good to go (adjective): Ready, everything is satisfactory or a situation is OK.

Gouge: (noun): Information or instructions, written or verbal.

Grand Old Man of the Marine Corps (noun): Nickname for Archibald Henderson, the fifth Commandant of the Marine Corps for 54 years. Also used to refer to the oldest Marine in the unit.

Green gear (noun): Can refer to encrypted radios, distinct from the "black gear" of unencrypted radios.

Green Side / Blue Side (noun): Refers to Navy Corpsmen who are assigned to and work directly with Marine units, often in combat environments.

Green Weenie (noun): Symbolic reference to Marine Corps bureaucracy or unfavorable circumstances.

Grinder (noun): Parade ground. See "Parade Deck."

Grog bowl (noun): Mess night tradition.

Ground pounder (noun): Infantry.

Grunt (noun): Marine infantryman.

Grunt proof (adjective): So basic even boots can't break.

Guadalcanal (noun): First major WWII offensive against Japan; Marines fought six months under brutal conditions.

Gucci gear (noun): High-end, often non-standard issue or luxury items worn by Marines.

Guidon (noun): Rectangular flag, typically carried by smaller units like platoons or companies.

Gun bunny (noun): Artillery Marine.

Gundeck (verb): To falsify information, take shortcuts or avoid tasks while falsely documenting compliance.

Gun doc (noun): Nickname for Navy Corpsman.

Gung Ho / Gungee (adjective): Extreme enthusiasm and dedication, often to the point of being overly zealous.

Gun line (noun): The position or area where artillery pieces are emplaced and ready to fire.

Gun pit (noun): Defensive position for crew-served weapons. Less formal reference to Rifle Range Pits where Marines operate targets for rifle range qualifications.

Gun truck (noun): A modified tactical vehicle equipped with heavy weapons like machine guns and sometimes armor plating, primarily for escort and protection duties.

Gunner (noun): A non-technical Chief Warrant Officer (0306 MOS) focused on weapons systems and tactics, or an enlisted MOS (0331) specializing in the tactical employment of machine guns.

Gunny (noun): Nickname for Gunnery Sergeant.

Gut truck (noun): A mobile food truck that provides refreshments and snacks to Marines where access to traditional dining facilities is limited. Also called a "Roach Coach."

Gyrene (noun): Slang term used to refer to Marines, believed to have originated as a derogatory nickname, but was ultimately embraced by Marines as an affectionate and self-descriptive term.

- H -

Half-mast (noun): Flag is lowered to a position halfway up the flagpole signifying mourning or respect, typically upon the death of a prominent individual or in times of national tragedy.

Hard charger (noun): A Marine who consistently performs exceptionally well, demonstrating both high levels of motivation and a strong work ethic, often exceeding expected standards.

Hard site (noun): Fortified structure or position designed to withstand attacks.

Hat (noun): Drill instructor.

Hatch (noun): Door or doorway.

Haze (verb): Extra-legal "toughening."

Head (noun): Bathroom.

Head call (noun): Bathroom break.

Heat casualty (noun): Overheated Marine.

Heat tab (noun): Ration heater.

Hero shot (noun): Pose with weapon.

Hesco (noun): Barrier baskets.

High and tight (noun): Classic haircut.

High drag (adjective): Clunky.

Hog's tooth (noun): Sniper 7.62 round.

Hollywood (adjective): All show.

Hollywood Marine (noun): MCRD San Diego grad.

Hollywood Parade Deck (noun): MCRD SD parade ground.

Hollywood shower (noun): Long wasteful shower.

Honey bucket (noun): Field toilet.

Honor man (noun): Top recruit.

Hooch (noun): Field shack.

Horn (noun): Radio.

Horse collar (noun): Wounded carry strap.

Hot brass (noun): Spent shells.

House mouse (noun): Recruit DI assistant.

Humvee (noun): High Mobility Multipurpose Wheeled Vehicle (HMMWV).

Hump (verb): Field march.

Hunter of Gunmen (HOG) (noun): Marine who successfully completes Scout Sniper School and earns the 0317 MOS.

Hurricane deck (noun): Upper ship deck.

Hurry up and wait (expression): Military irony.

Hush puppy (noun): Suppressed pistol.

- I -

Inspector-Instructor (I&I) (noun): Refers to active duty Marines assigned to support Reserve units, specifically the Selected Marine Corps Reserve (SMCR).

IAR (noun): M27 Infantry Automatic Rifle.

Ice cream suit (noun): Refers to the Marine Corps Blue-White Dress uniform.

ID10T form (noun): Joke "idiot form" prank on clueless Marines.

If it ain't raining, we ain't training (expression): Marine saying on embracing hardship.

If you ain't cheating, you ain't trying (expression): Often repeated in Marine infantry humor.

Improvise, adapt and overcome (expression): The unofficial motto of the Marine Corps.

In country (expression): Deployed operational.

In the rear with the gear (expression): Slang term describing personnel or units primarily involved in support roles, away from the front lines and active combat.

In the stack (expression): Waiting to breach.

In-country (adverb): Deployed operationally.

Initial issue (noun): First gear from supply.

Ink stick (noun): Pen.

Intel (noun): Intelligence.

Irish pennant (noun): Loose thread, important to remove prior to uniform inspections.

Iron Mike (noun): Statues of tough Marines.

It don't mean nothin' (expression): Vietnam-era saying, still lingers in grunt culture.

Iwo Jima (noun): Sacred WWII USMC battle site.

- J -

Jack up (verb): Reprimand hard.

Jane Wayne Day (noun): Families see training.

Jarhead (noun): A common nickname for Marines. Exact origin is debated.

Java (noun): Coffee, constant grunt fuel.

Jesus nut (noun): Main rotor retaining nut or mast nut that holds the main rotor assembly onto the mast of certain helicopters.

JJ DID TIE BUCKLE (noun): Mnemonic for the 14 leadership traits: **J**ustice, **J**udgement, **D**ependability, **I**nitiative, **D**ecisiveness, **T**act, **I**ntegrity, **E**nthusiasm, **B**earing, **U**nselfishness, **C**ourage, **K**nowledge, **L**oyalty and **E**ndurance.

Jody (noun): A civilian who, while a Marine is deployed or away on duty, is potentially becoming romantically involved with the Marine's spouse or partner.

John Wayne / Dead Duke (noun): P-38 can opener, a small folding blade used to open canned rations.

Joint ops (noun): Operating with other services, common in USMC.

Joker (noun): Most commonly refers to the radio call sign for Company G, 2nd Battalion, 4th Marines, specifically during their deployments to Iraq in 2004 and Afghanistan in 2011-2012.

Juice (noun): Power, often electrical or informal influence.

Jungle boots (noun): Vent-holed boots designed for hot, wet areas.

Junk on the bunk (noun): Inspection layout on a rack.

- K -

K-Bay (noun): MCAS Kaneohe Bay.

K-pot (noun): Nickname for the Personnel Armor System for Ground Troops (PASGT) helmet.

KA-BAR (noun): Legendary fighting knife designed during World War II.

Keep your head on a swivel (expression): Maintain situational awareness.

Khe Sanh (noun): USMC siege in Vietnam.

KIA (noun): Killed in action.

Kill! (expression): Can mean "yes, I understand," "hell yeah," or "let's do this." Also as a half-joking version of hello.

Kill hat (noun): Aggressive DI.

Klick (noun): Kilometer.

Knee cap (verb): Sabotage.

Knowledge (noun): Book.

Knuckle dragger (noun): Derogatory label for infantry Marines, particularly those perceived as having a simple, brute-force approach to problem-solving.

Krypto (noun): Encryption gear.

Kuni (noun): Nickname for MCAS Iwakuni.

- L -

Ladderwell (noun): Stairs.

Lance Coolie / Lance Criminal / Lance Coconut (noun): Slang for Lance Corporals.

Lance Corporal mafia (noun): Lance Corporal peer network.

Lance Corporal Underground (noun): Source of all rumors.

LCAC (noun): Landing Craft, Air Cushion (LCAC), a high-speed, fully amphibious hovercraft used by the U.S. Navy and Marine Corps to transport troops, weapons, equipment and cargo from ship to shore.

Leatherneck (noun): Nickname for Marines, referring to the high leather collars they once wore.

Leather Personnel Carriers (LPCs) (noun): Boots.

Leave (noun): Official vacation.

Lejeune (noun): Camp Lejeune.

Liberty (noun): Authorized free time.

Liberty risk (noun): Trouble-prone Marine.

Lifer (noun): Career Marine.

Lifer juice (noun): Coffee.

Light up (verb): To fire on an enemy.

LIMA Charlie (expression): Loud & clear, comms speak.

Lipstick Lieutenant (noun): Nickname for a Chief Warrant Officer 5.

Lit up (adjective): Shot or verbally torched.

Little bird (noun): Helos used by USMC aviation / SOCOM.

Live fire (noun): Actual ammo.

Living the dream (expression): Sarcastic morale phrase.

Load out (noun): Gear packed.

Load plan (noun): Document showing how gear is arranged for transport.

Lock and load or Lock and cock (interjection): To prepare a weapon to fire; also to get ready for action.

Lock it up (verb): Be quiet.

Lock on (verb): Focus.

Locked, cocked and ready to rock (expression): Fully prepared.

Log pack (noun): Logistics convoy.

Loggie (noun): Logistics Marine.

Long gun (noun): Sniper.

Lost lieutenant (expression): Joking reference to a new officer.

Loudspeaker ops (noun): Broadcasts to locals.

Low crawl (verb): Flat movement.

LT (noun): Short for Lieutenant.

LZ (noun): Landing zone.

- M -

M4 (noun): Standard carbine rifle.

Ma Deuce (noun): .50 cal machine gun.

Maggie's Drawers (noun): A red flag on the rifle range, signaling a miss.

Maggot (noun): Boot camp recruit.

MAGTF (acronym): Marine Air-Ground Task Force.

Main effort (noun): The principal task or objective that a commander focuses their resources and efforts on to achieve a specific mission

Mainside (noun): The main part of a Marine Corps base, often where the majority of facilities like the movie theater, bowling alley and PX (post exchange) are located.

Make a hole (command): Clear a path.

Mameluke Sword (noun): Officers carry the Mameluke Sword, which was originally given to Lieutenant Presley O'Bannon in 1805 by a Mameluke chieftain in North Africa.

Mandatory fun (noun): Planned social events that Marines are required to attend, often with the expectation of enjoying themselves.

Manpack (noun): Portable, tactical radios like the L3Harris Falcon IV family of radios.

Marine House (noun): Living quarters for Marines on Embassy Duty.

Mast (noun): CO's disciplinary hearing.

MCT (noun): Marine Combat Training for non-infantry after boot camp.

MEF (noun): Marine Expeditionary Force.

MEFEX (noun): MEF-level exercise.

Meritorious Mast (noun): A formal way for a commander to recognize and commend an enlisted Marine for outstanding performance and dedication to duty that goes above and beyond normal expectations.

Mess deck (noun): Ship dining, key to Marine ship deployments.

Mess dress (noun): Formal evening uniform for balls and high ceremonies.

Mess night (noun): Marine formal dinner.

MEU (noun): Marine Expeditionary Unit.

Mike (noun): Minute in comms speak.

Mike boat (noun): Refers to an LCM-8 (Landing Craft Mechanized, Mark 8).

Mike-mike (noun): Millimeter slang.

Military Left (noun): Used sarcastically when giving orders when a subordinate turns to the right instead of left.

Military Right (noun): Used sarcastically when giving orders when a subordinate turns to the left instead of right.

Moon dust (noun): A fine, powdery sand found in certain regions, particularly in southern Afghanistan, that can be problematic for equipment and personnel.

Moon beam (noun): Flashlight.\

Moon floss (noun): Toilet paper.

MOPP (noun): Mission-Oriented Protective Posture for CBRN threats.

Morning Colors (noun): The daily ceremony of raising the national flag at 08:00.

MOS (noun): Military Occupational Specialty.

Motard (noun): A slang term used to describe a Marine who is excessively enthusiastic or motivated, often to the point of being perceived as annoying or overbearing.

Motivated / Moto (adjective): Fired up.\

Motivated, dedicated and a little bit medicated (expression): Marine self-deprecating joke.

Motivator (noun): Very fired-up Marine.

Motor T (noun): Motor transport.

Mount out (verb): The process of assembling and deploying equipment and personnel for a mission, or the act of attaching medals and ribbons to a uniform.

Mount up (verb): A call to action, signaling the beginning of a movement or operation, often involving vehicles or mounted personnel.

MRE (noun): Meals Ready-to-Eat.

MSG (noun): Marine Security Guard, Embassy Guard.

MSR (noun): Main Supply Route.

Mustang (noun): Slang term used to refer to a Marine commissioned officer who began their career as an enlisted service member.

Muzzle awareness or Muzzle discipline (noun): Emphasizes the importance of always being aware of where the muzzle of a weapon is pointed and ensuring it is pointed in a safe direction.

- N -

Nap of the Earth (expression): Refers to a low-altitude flight technique used to minimize detection by enemy radar and other sensors.

NAVMC (noun): Marine Corps directive.

NCO (noun): Non-commissioned Officer.

NCO club (noun): Club for NCOs on base.

NCO Sword (noun): Adopted in 1859, the NCO Sword is carried by Marine NCOs and Staff NCOs. Used for ceremonial purposes, the M1859 NCO Sword was bestowed by the 6th Commandant, Col. John Harris, in recognition of their leadership in combat.

Negative (expression): "No."

New Breed (noun): Modern Marines.

New Corps (noun): Joking reference to softer New Corps vs. Old Corps.

Ninety-six (noun): Four-day liberty, 96 hours.

Ninja punch (noun): Slang for NJP.

NJP (noun): Non-Judicial Punishment. Often referred to as "Office Hours," is a process for handling minor misconduct without resorting to a court-martial. It's a disciplinary tool used by commanders to maintain order and discipline within their units.

No impact, no idea (expression): Didn't hit, clueless.

Non-hacker (noun): Individual who cannot perform a task.

Non-rate (noun): Marine below NCO.

No shit, there I was ... (expression): Start of Marine war stories.

No slack (expression): No tolerance.

Nut to butt (expression): Standing close in line.

- O -

O-Club (noun): Officers 'club.

O-course (noun): Obstacle course.

OCS (noun): Officer Candidate School.

O-Dark Thirty, Zero-Dark Thirty (noun): Refers to a very early and unspecified time in the morning, between midnight and sunrise.

Office Hours (noun): Administrative ceremony where legal, disciplinary, and other matters (such as praise, special requests, etc.) are attended, designed to dramatize praise and admonition, in a dignified, disciplined manner, out of the ordinary routine. See "NJP" and "Captain's Mast."

OIC (noun): Officer in Charge.

Oki (noun): Okinawa.

Old Breed (noun): WWII-era or earlier Marines.

Old Corps (noun): "Back when things were harder."

Old Man (noun): CO nickname, not used to his face.

Old Salt (noun): A veteran Marine, especially one who has served many years and has extensive experience at sea.

Oldest to youngest cake slice (verb): Birthday tradition.

On line (expression): Shoulder-to-shoulder formation.

On station (expression): In assigned area.

Once a Marine, Always a Marine (expression): Motto expressing lifelong identity.

Ooh-rah (interjection): Used to express enthusiasm. "Rah" is a shortened version and "Errrr" is an even more shortened version of "Rah." Also, see "Yut."

OP (noun): Observation Post.

Oscar Mike (expression): On the move.

Osprey (noun): Marine MV-22 aircraft.

Over the hill (expression): UA slang.

Over-penetration (noun): Bullet through target.

Overhead (noun): Ceiling.

Overwatch (noun): Covering fire.

- P -

Pack out (verb): Typically refers to the process of preparing and packing gear for a move, whether it's a Personal Change of Station (PCS) or a deployment.

Pain is weakness leaving the body (expression): PT motto.

Parade Deck (noun): Area set aside for the conduct of parades, drill, and ceremonies, often paved or well-maintained lawn. See "Grinder."

Parade rest (noun): A position of relaxation that can be assumed from the position of attention.

Pass in review (verb): A ceremonial event where a new commander inspects troops, often during a change of command or a graduation ceremony.

Pass the word (verb): Spread info.

Passageway (noun): A hallway or corridor.

Passed over (adjective): Having failed selection for the next higher rank for SNCOs and officers.

Pay grade (noun): Pay scale associated with rank and categorized as enlisted (E-1 to E-9), warrant officer (W-1 to W-5), and officer (O-1 to O-10).

PCS (noun): Permanent Change of Station.

PFC (noun): Private First Class.

PFT (noun): Physical Fitness Test.

Physical Conditioning Platoon (PCP) (noun): Program to help overweight recruits meet the physical standards for Marine Corps boot camp. Derisively called Pork Chop Platoon.

Pick up (verb): Start a new training cycle.

Professionally Instructed Gunman (PIG) (noun): Marine undergoing training to become a scout sniper, but not yet fully qualified. PIG is also a nickname for an M60 machine gun.

Pig egg (noun): 40mm grenade slang.

PMI (noun): Primary Marksmanship Instructor.

Pisscutter (noun): Nickname for soft green garrison cap.

Pit, The (noun): Physical fitness area in boot camp, often sandy.

Pizza Box (noun): Marksman badge, nicknamed for its square shape.

Plank owner (noun): Founding member of new unit.

POG, Pogue (noun): Non-infantry Marine.

Pogey bait (noun): Snacks.

Pogey rope (noun): French Fourragère.

Police (verb): To pick up items (such as litter or expended ammunition casings), to return an area to a natural state, or to correct another Marine.

Port (noun): Toward the left-hand side of the ship when facing forward.

Pos (noun): Position in comms speak. Pronounced "paws."

Position improvement (noun): Dig in defenses.

Power projection (noun): The ability to deploy and sustain military forces in and from dispersed locations to respond to crises, deter potential adversaries, and enhance regional stability.

Prep fire (noun): The initial stage of a fire mission where adjustments are made to ensure accurate fire, often involving a forward observer (FO) calling for spotting rounds and corrections before a full "fire for effect".

Pros & Cons (noun): Contraction of "Proficiency and Conduct marks," a numeric system for evaluating enlisted Marines.

PT (noun): Physical training.

PT gear (noun): Workout clothes.

PT stud (noun): Top performer.

Pugil Sticks (noun): Padded poles used as part of the Marine Corps Martial Arts Program (MCMAP) to simulate bayonet combat and teach close-quarters fighting techniques.

Pull Butts (verb): Mark targets on range.

PX (noun): Post Exchange.

- Q -

Quadcon (noun): Shipping container.

Quarterdeck (noun): Upper deck on a ship used by officers and for ceremonies; also refers to area in squad bay where recruits undergo incentive training (IT), (verb) "quarterdecking."

Quarters (noun): Barracks.

QRF (noun): Quick Reaction Force. A rapidly deployable, specialized unit designed to respond to a variety of crises and threats.

- R -

Rack (noun): Bed.

Rack out (verb): Go to sleep.

Rack time (noun): Time for sleep.

Rah (interjection): Short form of "Ooh-rah."

Rah motivator (noun): Overly moto Marine.

Range card (noun): A visual representation of a weapon's assigned sector of fire, used to assist in target acquisition and fire control.

Range qual (verb): Qualify on range.

Ratfuck (verb): Raid rations.

Rattle Battle (noun): Refers to the National Trophy Infantry Team Match (NTIT), a unique shooting competition.

Rear area (noun): Behind front lines.

Rear echelon (noun): Support troops.

Rear party (noun): A small number of Marines that stay "in country" for a longer period than the majority of the unit.

Recon (noun): Reconnaissance.

Red Patcher (noun): Landing support Marines.

Red team (noun): Simulated enemy.

Reenlistment bonus (noun): Cash to stay in.

Relief in place (RIP) (noun): Swap units.

Remain overnight (RON) (verb): Stay in place.

Remington Raider (noun): Marine admin clerk.

Request Mast (verb): Appealing to increasingly higher links in the chain of command in order to seek satisfaction for a grievance the requester feels was not adequately handled at a lower level.

Retrograde (verb): Withdraw.

Re-up (verb): Reenlist.

Reveille (noun): Morning bugle.

Roach coach (noun): Civilian food truck. Also called a Gut Truck.

Rock, The (noun): Okinawa.

Roger (expression): Acknowledged, comms talk.

Roger that (expression): Acknowledged. Same as "roger."

Run and gun (verb): Aggressive shoot & move.

- S -

SACO (noun): Substance Abuse Control Officer who provides substance abuse education/prevention, urinalysis screening and assistance to the Commander on substance abuse related matters.

SAFE (acronym): Used in water survival training. SAFE (**S**low Easy Movements, **A**pply Natural Buoyancy, **F**ull Lung Inflation, **E**xtreme Relaxation)

Salty (adjective): Experienced Marine. Also, "Salty Dog"

Sand fleas (noun): A common nuisance for Marines stationed at Marine Corps Recruit Depot Parris Island, South Carolina.

Sand table (noun): Sand model for visualizing battlefield.

SAW (noun): Squad Automatic Weapon.

Say again (phrase): Request to repeat a statement, question, or order, especially over a radio.

Schmuckatelli, Joe (noun): Generic Marine.

School circle (noun): Marines seated around instructor.

Scullery (noun): Place where dishes are washed.

Screw the pooch (verb): Screw up.

Scribe (noun): Recruit who writes notes and schedules for DIs.

Scuttlebutt (noun): Rumor, gossip.

Scuzz brush (noun): Cleaning brush.

Sea Bag (noun): A large, cylindrical bag used to store and transport personal items, uniforms and gear

Secret Squirrel (noun/adjective): Related to intelligence personnel or clandestine, covert, classified or confidential activity or information.

Sea Going Bellhop (noun): Marine serving aboard Navy ships as sentries, security, orderlies, honor guards for special occasions, and the nucleus of the ship's landing party.

Sea Lawyer (noun): One who questions or argues about orders.

Sea Bag (noun): Marine's big canvas bag.

Secure (verb): Close, lock up, take care of.

Semper Fidelis / Semper Fi (expression): Latin for "Always Faithful," the official motto of the Marine Corps.

Semper Fu (noun): Marine Corps martial arts.

Semper Gumby (expression): Always flexible.

Semper I (expression): Marine who acts only for self.

Send in the Marines (expression): Cultural cliché because Marines are often first used in crises.

S/F (abbreviation): Semper Fidelis.

Shellback (noun): Marine who has taken part in the crossing of the line ceremony or crossing the equator ceremony while on a naval vessel.

Ship Over (verb): To reenlist.

Shitbird (noun): Undisciplined or unsat Marine.

Shit can (noun, verb): Trash can, throw out.

Shit-hot (adjective): Term used to notify something as exceptional or very good.

Shit on a shingle (noun): Chipped beef on toast.

Short (adjective): Nearing EAS or redeployment.

Short round (noun): Ordnance landing too short of the target.

Short timer (noun): Nearing EAS.

Shower Shoes (noun): Recruit-issued flip-flops.

Sick bay (noun): A room or building set aside for the treatment or accommodation of the sick within a military base or on board a ship.

Sick Call (noun): Daily period when routine ailments are treated at sick bay.

Side straddle hop (noun): Jumping jacks.

Silkies (noun): Marine PT shorts.

Silver bullet (noun): Rectal thermometer.

SITFU (acronym): Suck it the fuck up.

SITREP (noun): Situation report.

Six (noun): Your back.

Skate (verb/adjective): Slacking off or easy duty.

Skipper (noun): CO nickname.

Skivvies (noun): Underwear.

Skivvy shirt (noun): White T-shirt.

Slack man (noun): 2nd in breaching stack.

Slick sleeve (noun): A Private. No rank insignia.

Slop chute (noun): Bar, PX restaurant or E-club.

SMEAC (noun): Five paragraph order format for operation orders — SMEAC (**S**ituation, **M**ission, **E**xecution, **A**dmin & Logistics, **C**ommand & Signal). See Five Paragraph Order entry.

Smoke (verb): Intense PT punishment.

Smoke check (verb): Inspect hard.

Smoker (noun): Boxing match.

Smokey bear (noun): DI campaign cover.

Smoke and Joke, Smoking and Joking (adjective): Acting unproductively.

Smoking lamp (noun): Shipboard tradition indicates permission to smoke or to stop smoking.

SNAFU (noun): Situation normal, all fucked up.

Snap in (verb): Conduct sighting in or aiming exercises with an unloaded weapon.

Snap link (noun): Carabiner.

SNCO (noun): Staff Non-Commissioned Officer (E-6 to E-9).

Snivel gear (noun): Cold weather gea

Snivel list (noun): List of complainers.

Snowflake (noun): Soft Marine.

Snowshoe (noun): Mortar base plate.

Soft cover (noun): Garrison cap.

Soft target (noun): Lightly defended.

Soissons & Champagne (noun): WWI battles earned French Fourragère.

Sound off (verb): Loudly respond.

Soup locker (noun): Mouth or jaw.

Soup sandwich (noun): Complete mess.

Sour (expression): Not working / bad.

Space available (noun): Open room on flights.

Spin dry (verb): Quick fix.

Spotter (noun): Sniper assistant

Spring butt (noun): Dummy target.

Squad automatic weapon (noun): M249 (SAW).

Squadbay (noun): Living quarters with open rooms and shared head.

Squad leader (noun): A crucial role, typically held by a Sergeant (E-5), responsible for the discipline, training, and welfare of their squad, as well as the tactical employment of their Marines in combat.

Squared away (adjective): Neat and competent Marine.

Squid (noun): Nickname for Navy sailor. See "swabbie."

SRB (noun): Service Record Book, an administrative record of an enlisted Marine's personal information, promotions, postings, deployments, punishments, and emergency data.

Stack (verb): Line up close for entry.

Stack up (verb): Get on door for breach.

Stand by (expression): Wait for orders.

Stand by to stand by (expression): Classic "hurry up & wait."

Stand to (verb): Defend at dawn/dusk

Starboard (noun): The right side of a ship or boat; toward the right-hand side of a vessel facing forwar

Stay frosty (expression): To remain alert, focused, and calm under pressure, especially during combat or high-stress situations.

Stumps, The (noun): Nickname for the Marine Corps Air Ground Combat Center (MCAGCC) at Twentynine Palms, California.

Suck (noun): Mouth. "The Suck" refers to a miserable situation, hardship or the Marine Corps.

Suzy Rottencrotch (noun): Generic girlfriend.

Swab (verb): Mop.

Swabbie (noun): Nickname for Navy Sailor. See "squid."

Sweat locker (noun): Wait nervously.

Swim qual (noun): USMC swim test.

Swingin 'with the wing (expression): Duty with USMC aviation.

Swinging Dick (noun): Generic Marine.

Swoop (verb): Dash home on libo.

Swoop circle (noun): Place to hook a ride from a car-owning Marine.

- T -

Tactically acquire (verb): To steal something.

TAD (noun): Temporary Additional Duty.

Tango (noun): Target/enemy.

Tango Mike (expression): Thanks much.

Taps (noun): Bugle at night/funeral.

Tarawa (noun): Bloody WWII USMC assault.

TBS (noun): The Basic School at Quantico.

Term of endearment (noun): Sarcastic NCO sweet-talk.

Terminal Lance (noun): LCpl stuck at E-3.

That's above my pay grade (expression): Not my lane.

The Few. The proud. (expression): Recruiting motto.

The only easy day was yesterday (expression): Embrace challenge.

Three hop drop (noun): Gear drops w/light jostle.

Three-block war (noun): Marine urban ops concept.

Through and through (noun): Bullet in/out wound.

Thumper, Thump Gun (noun): Grenade launcher.

TIC (noun): Troops in contact.

Tip of the spear (expression): Term for a unit that enters enemy territory first.

Tombstone courage (noun): Reckless bravery.

Top (noun): Nickname for a Master Sergeant or Master Gunnery Sergeant, inappropriate to use without permission.

Topside (noun): Upstairs, on deck.

Tracers work both ways (expression): Enemy sees tracers.

Tracks (noun): Armored vehicles.

Transient barracks (noun): Short-term lodging.

Transients (noun): Refers to Marines who are temporarily assigned to a unit or location for a specific purpose, such as training, undergoing medical treatment, or attending a course.

Transpo (noun): Motor Transport.

Trash detail (noun): Refers to the responsibility of Marines to maintain cleanliness within their unit's area, including barracks and surrounding grounds.

Trench broom (noun): Refers to the Winchester Model 1897 shotgun, specifically when modified for trench warfare. It earned this nickname due to its use in close-quarters combat within trenches during World War I.

Tripoli (noun): Refers to 1805 Derna during the Barbary Wars.

Trunk monkey (noun): Refers to a nickname for a Marine who provides rear security, often in a vehicle during convoys.

Tun Tavern (noun): Founding site of the Corps on 10 November 1775 in Philadelphia.

Turn to (verb): Start working.

Two is one, one is none (expression): Marine redundancy rule. Always carry a backup.

Two-block (verb): Hoist a flag or pennant to the peak, truck, or yardarm of a staff; or a tie with the knot positioned exactly in the gap of a collar of a buttoned shirt.

- U -

UA (noun): Unauthorized absence.

UCMJ (Uniform Code of Military Justice) (noun): Military legal code.

Un-ass (verb): Get out quick.

Un-fuck (verb): To fix something or correct a deficiency.

Under arms (adjective): Carrying a weapon.

Under the table of organization (expression): Extra bodies not on books.

Underway (adjective): To depart or to start a process for an objective.

Undress blues (noun): Refers to the Blue Dress "D" or "C" uniforms.

Uniform of the day (noun): The designated uniform that Marines are required to wear on a given day, as determined by their unit commander.

Unit diary (noun): A comprehensive, official record of a unit's activities and personnel changes.

Unit sweep (noun): Barracks clean.

UNQ (adjective): Unqualified. Pronounced "unk."

Unsat (adjective): Unsatisfactory.

USMC (noun): United States Marine Corps.

Utilities (noun): Camouflage uniform.

- V -

V-ring (noun): Center of silhouette.

Veg (verb): Zone out.

VIC (noun): Vehicle.

Vicinity (noun): Nearby.

Victor unit (noun): Radio ground callsign. (NATO phonetic "V" often designates ground forces).

Voluntary recall (noun): Reservists activated.

Voluntold (verb): "Volunteer" without choice.

- W -

Walking John (noun): Nickname for a Marine marching in dress blues uniform that appeared on World War I-era recruiting posters.

War belt (noun): A web belt used to carry canteens in pouches and other miscellaneous equipment.

Watch (noun): Guard duty period.

Watch your lane (expression): Mind your own.

Water Buffalo (noun): Trailer-mounted water tank.

Weapons tight (expression): A control order indicating that weapons systems can only be fired at targets positively identified as enemy or hostile.

Weapons free (expression): Signifies that weapons can be fired at any target not positively identified as friendly.

Welcome to the suck (expression): A resigned way to accept that things are going to be tough.

Wet down (noun): Party thrown by a newly promoted Marine, traditionally pays for the drinks.

What happens in the field stays in the field (expression): Field antics stay in the field.

Whiskey Delta (expression): Weak dick, often playful but still sharp ribbing.

Whiskey locker (noun): Supply locker.

Whiskey Tango (expression): White trash.

Whiskey Tango Foxtrot (expression): WTF, What the fuck.

Whiz quiz (noun): Urinalysis.

Who's your buddy? (expression): Watch your fellow Marine.

Willie Pete (noun): White phosphorus.

Winger (noun): Aviation Marine.

Wing Wiper (noun): Aviation Marine, usually in maintenance, not a pilot.

WM (noun): Woman Marine. Now discouraged in official usage.

Wook (noun): A female Marine.

Wooly Pully (noun): Green wool sweater.

Work your bolt (verb): To resort to special measures to attain a particular end. From the action of racking a rifle's bolt to clear a stoppage.

Working party (noun): A group of Marines assigned to perform tasks outside of their primary duty, often involving manual labor.

- X -

X-fill (noun): Exfiltration, the process of safely withdrawing personnel or units from a hostile, dangerous, or operational area, often under the threat of enemy forces.

XO (noun): Executive officer.

- Y -

Yardbird (noun): Derogatory term for a Marine who avoids work or shirks responsibilities.

YAT-YAS (acronym): You ain't track, you ain't shit. Motto and a rallying cry for Amphibious Assault Vehicle (AAV) Marines.

Yellow Footprints (noun): The very first position of attention a recruit assumes on Marine Corps soil, symbolizing the start of the transformation from civilian to United States Marine.

Yut or Yut Yut (interjection): Yelling Unnecessary Things. Motivational saying similar to Oo-rah.

- Z -

Zap number (noun): A method of personal identification used in combat situations, typically written on gear like body armor.

Zero (verb / noun): The process of adjusting the sights of a weapon, specifically the elevation and windage settings, to ensure accurate bullet impact on a target at a specific range. Also used to refer to officers, derived from the "o" in officer.

Zero-dark thirty, O-dark thirty (noun): Refers to a very early and unspecified time in the morning, between midnight and sunrise.

Zero in (verb): Align sights. See "Zero."

Zeroed out (verb): Finished responsibilities.

Zipperhead (noun): Vietnam-era enemy.

Zone recon / Zone reconnaissance (noun): A specific type of reconnaissance mission focused on gathering detailed information about a defined area.

Zoomie (noun): Nickname for an Air Force pilot.

Numeric Terms

03 Hump-a-Lot (noun): Pejorative used by support Marines for infantry. (03 = infantry MOS series)

1st Civ Div (expression): 1st Civilian Division. Civilian life. Applied to Marines facing discharge or retirement. Also referred to as 1st Couch Company and Camp Living Room.

360 (noun): Full circle security.

48, 72, 96 (noun): Liberty time blocks in hours (48 = 2 days, 72 = 3 days, 96 = 4 days).

4th Battalion (expression): Pejorative for soft Marines; specific PI joke. (4th Battalion is traditionally female recruit training).

4th Marine Dimension (expression): Derogatory for Marine Forces Reserve division.

5.56 hickey (noun): A scar or blister resulting from a burn suffered due to hot brass.

782 Gear (noun): Term for issued field gear. Refers to DD form signed when issued gear. Also called "Deuce" gear.

8th & I (noun): Marine Barracks Washington, D.C. — oldest active post in the Corps.

Appendix 1: Pronunciation Guide

Here's a quick rundown to keep civilians or new boots from butchering Marine speak:

Word or Phrase	How Marines Say It
FMF	"Eff-Em-Eff." Fleet Marine Force
Gung Ho	"Gung" rhymes with "rung," not "gong."
Gyrene	"Jye-reen"
Hooch	"Hooch" like "pooch."
Ma Deuce	"Mah Doose" (the .50 cal)
Oorah	"OO-rah!" (short, guttural, from the chest)
Semper Fi	"SEMP-er Fye" (not "fee")
SITREP	"SIT-rep," short for Situation Report.

Word or Phrase	How Marines Say It
SNAFU	"SNA-foo." Situation Normal, All Fucked Up
Tun Tavern	"Tun" rhymes with "gun."
UA	"You-Ay," Unauthorized Absence.

Appendix 2: Marine-to-Civilian Decoder Chart

Because even the most motivated civilians have no idea what half this means — and boots sometimes need a primer.

Marine Term	Civilian Translation
Bird	Aircraft
Blouse	Camouflage shirt top
Bulkhead	Wall
Cammies	Camouflage uniform
Cover	Hat
Deck	Floor
Field day	Thorough cleaning session
Fobbit	Someone who stays on base

Marine Term	Civilian Translation
Grunt	Infantry Marine
Gunny / Top	Gunnery Sergeant / Master Sergeant
Head	Bathroom
Hump	March under load
Liberty	Free time
Lifer	Career Marine
Motivated / Moto	Enthusiastic, gung-ho
NJP / Ninja Punch	Non-judicial punishment
Old Corps	Marines of previous generations
Passageway	Hallway
POG / Pogue	Non-infantry personnel
PT	Physical training, working out
Rack	Bed
Scuttlebutt	Gossip (originally a water barrel)
Short timer	Almost done with tour or contract
SITREP	Situation Report
UA	Unauthorized Absence (< 30 days)
Zero-dark thirty	Very early morning

Appendix 3: Naval Terms Every Marine Should Know

From their birth on 10 November 1775, at Tun Tavern in Philadelphia, Marines were created as naval infantry — warriors of both sea and shore. Tasked to serve aboard ships of the fledgling Continental Navy, Marines stormed decks, manned cannons and carried the fight ashore in the earliest amphibious raids.

Over the past two-plus centuries, this bond only deepened: from boarding pirate vessels in the Caribbean and crossing the decks of ironclads in the Civil War, to launching from Navy ships onto Pacific beaches in WWII and into the sands of Iraq and Afghanistan.

The language of the sea — "bulkhead," "deck," "hatch," "scuttlebutt" — remains woven into the fabric of Marine life, a proud testament that while Marines may fight most fiercely on land, they have always been forged in a naval tradition, ready to strike from the sea.

Term	Meaning
Aft	Toward the stern of a ship.
Amidships	Middle of the ship, side to side.

Term	Meaning
Ashore	Any place outside a naval or Marine Corps installation.
As You Were	Resume former activity or position.
Aye-aye	An acknowledgment of orders. Means an individual understands and will comply with orders.
Belay	To make fast or secure, as in "belay the line."
Bulkhead	Wall of a ship or building.
Carry On	Resume what you were doing; as you were.
Deck	Floor on a ship (or by tradition, any floor).
Ensign	National flag flown on a ship, or Navy officer rank.
Fore / Forward	Toward the front (bow) of the ship.
Galley	Ship's kitchen.
Gangway	Passageway or opening in the ship's side; also "Gang way!" means "move out of the way!"
Hatch	Door or opening in the deck or bulkhead.
Head	Bathroom.
Ladderwell	Stairs.
Mess Deck	Dining area on ship.
Port	Left side of the ship when facing forward.

Term	Meaning
Quarterdeck	An upper deck on a ship that is used by officers; specifically for ceremonial use.
Scuttlebutt	Drinking fountain; also means rumor or gossip.
Sea Bag	Duffel bag for uniforms and personal gear.
Secure	Stop work or lock down.
Sick Bay	Ship's medical clinic.
Starboard	Right side of the ship when facing forward.
Topside	Upstairs, on deck.

Appendix 4: NATO Phonetic Alphabet

The NATO Phonetic Alphabet was developed to ensure clear, unambiguous communication over radio and telephone — especially in noisy, high-stress environments. Each letter of the alphabet is represented by a distinct word to prevent confusion between similar-sounding letters such as "B" and "D" or "M" and "N." It's a standardized language that ensures orders, coordinates and callsigns are understood the first time — no need to repeat, and no room for mistakes.

Letter	Code Word	Letter	Code Word
A	Alpha	N	November
B	Bravo	O	Oscar
C	Charlie	P	Papa
D	Delta	Q	Quebec
E	Echo	R	Romeo
F	Foxtrot	S	Sierra
G	Golf	T	Tango

Letter	Code Word	Letter	Code Word
H	Hotel	U	Uniform
I	India	V	Victor
J	Juliett	W	Whiskey
K	Kilo	X	X-ray
L	Lima	Y	Yankee
M	Mike	Z	Zulu

Appendix 5: Radio Brevity Codes

Radio brevity codes are standardized, concise words or phrases used by Marines and other military forces to streamline radio communication and reduce the chance of confusion. These codes allow operators to transmit critical information quickly and clearly when every second counts. Brevity keeps the airwaves clear, minimizes chatter and ensures that everyone is on the same page. It's not just comms — it's combat efficiency, one word at a time.

Code	Meaning
Actual	Unit commander
All After / All Before	Repeat all after or before a specific word.
Affirmative	Yes.
Be Advised	Important information follows.
Break	Pause in transmission, message continues after.
Break-Break	Urgent interrupt, higher priority traffic follows.
Cease Fire	Stop all firing immediately.

Code	Meaning
Check Fire	Stop firing temporarily.
Copy	Same as Roger; acknowledged receipt.
Correction	Cancels what came before and tells the listener to replace it with what follows.
Danger Close	Fire mission is close to friendly forces.
ETA	Estimated Time of Arrival.
FRAGO	Fragmentary Order (follow-on mission).
How Copy?	How do you hear me? Asking for clarity/readability.
I Say Again	I am repeating for clarity.
Loud and Clear / Lima Charlie	Response to How Copy: good signal.
Negative	No.
Oscar Mike	On the Move.
Out	This conversation is complete, no reply expected. (Never "Over and Out" together.)
Over	My transmission is complete, I await your reply.

Code	Meaning
Roger	Received and understood.
Say Again	Repeat your last transmission.
SITREP	Situation Report.
Splash	Rounds on target (artillery/mortar).
Standby	Wait — I'll get back to you.
Tango Mike	Thanks Much.
TIC	Troops in Contact (engaged with enemy).
Wait One	Very short pause (seconds).
Weapons Free	Engage any target not positively identified as friendly.
Weapons Hold	Do not fire except in self-defense.
Weapons Tight	Engage only positively identified hostile targets.
Wilco	Will comply (understood *and* will execute).

Appendix 6: Marine Corps Ranks & Insignia

ENLISTED	
E-1: Private (PVT)	No insignia
E-2: Private First Class (PFC)	
E-3: Lance Corporal (LCpl)	
E-4: Corporal (Cpl)	

ENLISTED (CONTINUED)

E-5: Sergeant (Sgt)	
E-6: Staff Sergeant (SSgt)	
E-7: Gunnery Sergeant (GySgt)	
E-8: Master Sergeant (MSgt)	

ENLISTED (CONTINUED)

E-8: First Sergeant (1st Sgt)

E-9: Master Gunnery Sergeant (MGySgt)

E-9: Sergeant Major (SgtMaj)

E-9: Sergeant Major of the Marine Corps (SMMC)

WARRANT OFFICERS

Warrant Officer 1 (WO1)
Red with gold background

Chief Warrant Officer 2 (CWO2)
Red with gold background

Chief Warrant Officer 3 (CWO3)
Red with silver background

Chief Warrant Officer 4 (CWO4)
Red with silver background

WARRANT OFFICERS (CONTINUED)

Chief Warrant Officer 5 (CW05) Red with silver background	
Chief Warrant Officer 2-5 (Gunner)	

COMMISSIONED OFFICERS

O-1: 2nd Lt. (2ndLt)
Gold bar

O-2: 1st Lt. (1stLt)
Silver bar

O-3: Captain (Capt)
Silver bars

O-4: Major (Maj)
Gold Oak Leaf

O-5: Lt. Colonel (LtCol)
Silver Oak Leaf

O-6: Colonel (Col)

COMMISSIONED OFFICERS (CONTINUED)

O-7: Brigadier General (BGen)	★
O-8: Major General (MajGen)	★★
O-9: Lieutenant General (LtGn)	★★★
O-10: General (Gen)	★★★★

Appendix 7: Key Marine Corps Dates & Battles

Date	Event
10 Nov. 1775	Birth of the Marine Corps. Continental Congress establishes the Corps at Tun Tavern, Philadelphia.
3 March 1776	First amphibious landing at New Providence, Bahamas. Marines seize British gunpowder stores.
April 1783	Continental Navy and Continental Marines disbanded following the American Revolution.
11 July 1798	Marine Corps re-established.
27 April 1805	Battle of Derna (Tripoli). "To the shores of Tripoli" — Marines raise the American flag in victory during the Barbary Wars.
13-14 Sep. 1814	Defense of Fort McHenry. Marines help repel the British, inspiring the Star-Spangled Banner.
March 1847	Battle of Veracruz (Mexican-American War). Major amphibious landing by Marines and Army.

Date	Event
13 Sep. 1847	Battle of Chapultepec. Marines storm Mexican fortress, earning the blood stripe tradition.
1-26 June 1918	Battle of Belleau Wood (WWI). Marines earn "Teufel Hunden" (Devil Dogs) nickname from Germans.
10 Nov. 1921	Marine Corps Birthday formalized. Commandant Lejeune issues order making 10 Nov the official birthday celebration.
1927 - 1933	Nicaragua campaigns. Legendary jungle fights under Chesty Puller.
7 Aug. 1942	Guadalcanal campaign begins (WWII). First major offensive against Japan.
19 Feb. 1945	Marines raise flag on Iwo Jima's Mt. Suribachi.
1 April - 22 June, 1945	Battle of Okinawa. Largest amphibious assault in the Pacific.
15 Sept. 1950	Battle of Inchon (Korean War). MacArthur's daring landing leads to Seoul recapture.
Nov-Dec 1950	Battle of Chosin Reservoir. Marines break out of Chinese encirclement; "Frozen Chosin."

Date	Event
8 March 1965	Marines land at Da Nang (Vietnam War). First U.S. ground combat troops in Vietnam.
30 Jan. 1968	Tet Offensive. Marines fight fierce urban battles at Hue City.
March 1991	Operation Desert Storm ends. Marines key in breaching Iraqi lines.
November 2004	Second Battle of Fallujah (Iraq). Fiercest urban combat since Hue.
2001 - 2021	Operation Enduring Freedom (Afghanistan).

Appendix 8: Historic Quotes From and About Marines

"That two battalions of Marines be raised consisting of one colonel, two lieutenant colonels, two majors and officers as usual in other regiments, that they consist of an equal number of privates with other battalions; that particular care be taken that no person be appointed to office or enlisted into said battalions, but such as are good seamen, or so acquainted with maritime affairs as to be able to serve to advantage by sea."

— RESOLUTION OF THE CONTINENTAL CONGRESS, 10 NOVEMBER 1775

"A ship without Marines is like a garment without buttons."

— ADMIRAL DAVID D. PORTER, USN, 1863

"The Marines have landed and have the situation well in hand."

— ATTRIBUTED TO MANY SOURCES AND POPULARIZED BY THE CORRESPONDENT RICHARD HARDING DAVIS DURING THE LATE NINETEENTH-CENTURY

"To our Marines fell the most difficult and dangerous portion of the defense by reason of our proximity to the great city wall and the main city gate ... The Marines acquitted themselves nobly."

— MR. EDWIN N. CONGER, U.S. MINISTER, IN COMMENDING THE MARINES FOR THE DEFENSE OF THE LEGATIONS AT PEKING, CHINA, IN 1900

"Your Marines having been under my command for nearly six months, I feel that I can give you a discriminating report as to their excellent standing with their brothers of the army and their general good conduct."

— GENERAL JOHN J. PERSHING, USA, IN A LETTER TO MAJOR GENERAL COMMANDANT GEORGE BARNETT, USMC, 10 NOVEMBER 1917

"I have only two out of my company and 20 out of some other company. We need support, but it is almost suicide to try to get it here as we are swept by machine gun fire and a constant barrage is on us. I have no one on my left and only a few on my right. I will hold."

— FIRST LIEUTENANT CLIFTON B. CATES, USMC,
96TH CO., SOISSONS, 19 JULY 1918.

"Come on, you sons of bitches, do you want to live forever?"

— GUNNERY SERGEANT DAN DALY,
BELLEAU WOOD, 1918

They (Women Marines) don't have a nickname, and they don't need one. They get their basic training in a Marine atmosphere, at a Marine Post. They inherit the traditions of the Marines. They are Marines."

— LIEUTENANT GENERAL THOMAS HOLCOMB,
USMC, 1943

"Casualties many; Percentage of dead not known; Combat efficiency; we are winning."

— COLONEL DAVID M. SHOUP, USMC,
TARAWA, 21 NOVEMBER 1943

"The raising of that flag on Suribachi means a Marine Corps for the next 500 years."

— JAMES FORRESTAL, SECRETARY OF THE NAVY, 23 FEBRUARY 1945

"Among the men who fought on Iwo Jima, uncommon valor was a common virtue."

— FLEET ADMIRAL CHESTER W. NIMITZ, USN, 16 MARCH 1945

"The bended knee is not a tradition of our Corps."

— GENERAL ALEXANDER A. VANDERGRIFT, USMC, TO THE SENATE NAVAL AFFAIRS COMMITTEE, 5 MAY 1946

"We're surrounded. That simplifies things."

— LIEUTENANT GENERAL LEWIS B. "CHESTY" PULLER, CHOSIN RESERVOIR, 1950

"I have just returned from visiting the Marines at the front, and there is not a finer fighting organization in the world."

— GENERAL DOUGLAS MACARTHUR, USA, OUTSKIRTS OF SEOUL, 21 SEPTEMBER 1950

"Once a Marine, always a Marine!"

— MASTER SERGEANT PAUL WOYSHNER, A 40-YEAR MARINE, IS CREDITED WITH ORIGINATING THIS EXPRESSION DURING A TAPROOM ARGUMENT WITH A DISCHARGED MARINE

"I can't say enough about the two Marine divisions. If I use words like brilliant, it would really be an under-description of the absolutely superb job they did in breaching the so-called impenetrable barrier ... Absolutely superb operation, a textbook, and I think it'll be studied for many, many years to come as the way to do it."

— GENERAL H. NORMAN SCHWARZKOPF, USA, RIYADH, SAUDI ARABIA, 27 FEBRUARY 1991

"Be polite, be professional, but have a plan to kill everybody you meet."

— GENERAL JAMES "MAD DOG" MATTIS

"No better friend, no worse enemy."

— ANCIENT ROMAN SAYING, ADOPTED BY MARINES, POPULARIZED BY GENERAL MATTIS

Appendix 9: Marine Corps Values

Honor
Honor guides Marines to exemplify the ultimate in ethical and moral behavior. Never lie, never cheat or steal; abide by an uncompromising code of integrity; respect human dignity and respect others. Honor compels Marines to act responsibly, to fulfill our obligations and to hold ourselves and others accountable for every action.

Courage
Courage is the mental, moral, and physical strength ingrained in Marines. It carries us through the challenges of combat and aids in overcoming fear. It is the inner strength that enables us to do what is right, to adhere to a higher standard of personal conduct, and to make tough decisions under stress and pressure.

Commitment
Commitment is the spirit of determination and dedication found in Marines. It leads to the highest order of discipline for individuals and units. It is the ingredient that enables constant dedication to Corps and country. It inspires the unrelenting determination to achieve victory in every endeavor.

Appendix 10: The Creed of the U.S. Marine

The Rifleman's Creed is accredited to Major General William H. Rupertus, USMC (Deceased) and still taught to Marines undergoing Basic Training at the Recruit Depots at Parris Island and San Diego. It was first published in the San Diego Marine Corps Chevron on 15 March 1942.

This is my rifle. There are many like it, but this one is mine.

My rifle is my best friend. It is my life. I must master it as I must master my life.

My rifle, without me, is useless. Without my rifle, I am useless. I must fire my rifle true. I must shoot straighter than my enemy who is trying to kill me. I must shoot him before he shoots me. I will ...

My rifle and myself know that what counts in this war is not the rounds we fire, the noise of our burst, nor the smoke we make. We know that it is the hits that count. We will hit ...

My rifle is human, even as I, because it is my life. Thus, I will learn it as a brother. I will learn its weaknesses, its strength, its parts, its accessories, its sights and its barrel. I will ever guard it against the ravages of weather and damage as I will ever guard my legs, my arms, my eyes and my heart against damage. I will keep my rifle clean and ready. We will become part of each other. We will ...

Before God, I swear this creed. My rifle and myself are the defenders of my country. We are the masters of our enemy. We are the saviors of my life.

So be it, until victory is America's and there is no enemy, but peace!

ABOUT THE AUTHOR

John Burke served in the United States Marine Corps from 1983 to 1989. As an Intelligence Specialist (0231) he served with 1st Marine Regiment and 1st Battalion, 9th Marines at Camp Pendleton, California, before guarding U.S. Embassies as a Marine Security Guard (MSG, 8151) in Guatemala City; Guatemala; Prague, Czechoslovakia; and Helsinki, Finland. After MSG duty, he served with VMFA-235 (The Death Angels) at MCAS Kaneohe Bay, Hawaii.

Following his time in uniform, John earned a B.A. in English (Summa Cum Laude) at the University of Southern Maine in his hometown of Portland, Maine, and went on to build a career in journalism, writing and corporate communications after relocating to Florida.

Semper Fidelis remains more than a motto for John — it's been the guiding compass of his life, and the spirit behind preserving the rich language and legacy of the Corps in *The Devil Dog Dictionary*.

www.ingramcontent.com/pod-product-compliance
Lightning Source LLC
Chambersburg PA
CBHW070639030426
42337CB00020B/4088